Theresa Löwenstein

Urban Garden USA. Community Gardening as a Tool of City Planning

GRIN Publishing

Bibliographic information published by the German National Library:

The German National Library lists this publication in the National Bibliography; detailed bibliographic data are available on the Internet at http://dnb.dnb.de .

Imprint:

Print and binding: Books on Demand GmbH, Norderstedt Germany
ISBN: 978-3-656-96295-3

This book at GRIN:

http://www.grin.com/en/e-book/299817/urban-garden-usa-community-gardening-as-a-tool-of-city-planning

Community Gardening as a Tool of City Planning

Research Paper
PA 525
Spring 2014

San Diego State University

Student: Theresa Hayessen

Date: San Diego, 05/05/2014

Contents

"In future ever more people will be living in cities rather than in rural areas. The city will therefore become the decisive place for the development of more sustainable ways of eating, living and moving. The city of the future should be a climate-friendly, pleasant place to live, where every care is taken to conserve our natural resources."

(Prinzessinengärten, 2014)

List of Figures

List of Abbreviations

CA	California
CG	Community Garden/s = Urban Gardens
LEED	Leadership in Energy and Environmental Design
SDCGN	San Diego Community Garden Network
U.S.	United States

1. The Origin of Urban Gardens

In modern times society in most U.S. cities changed to "bedroom communities" where people stay home, watch television and forget how to live in the cultural, urban or even village sense (Register, 2010). At the same time, however, issues such as global warming and sustainability gained attention, which led to the re-emergence of a movement within the city: Community Gardening.

The question to be answered is if Community Gardens/Urban Gardens (CG) are planned as a means to other objectives or an end in itself. If it is a means to other ends, the CG is only beneficial until the other aim is achieved. If not, the gardens serve a greater use than only to overcome crisis. This in turn would be an indicator that CG should be more recognized as a city-planning tool instead of decrease in times of peace and wealth. In order to find out what role CG play in todays urban planning and how it can contribute to improve urban conditions, I first have to illuminate the current problems in today's cities. After having a general overview on the present urban conditions I than focus on the historical and current development of CG in general and in particular in the United States (U.S.). Later I take a closer look on the general objectives behind the emergence of urban garden movements and the benefits that they contained in the past and present. Looking at recent prime examples of urban gardens in Berlin and San Diego will shed light on the goals behind and particular the benefits CG have on today's urban environments, communities and its residents. In the end I'll give some recommendations on how CG should be implemented in the field of city planning to improve the described urban conditions.

A. Challenges of Today's Urban Environments

Only a century ago the typical American community was a small town where citizens knew each other and could find everything for their daily needs within walking distances.

Nowadays more than half of all Americans live in anonymous metropolitan areas of over a million residents, where you could walk all day in order to cross the vast geographical space taken by the connected urban and suburban districts. In those new city constructs where only few people can cover the totality of their daily travels under a ten-minute walk, the sense of community has changed and the automobile became omnipresent (Calthorpe/Fulton, 2001). Through the dominance of cars the urban picture changed insofar as that car lanes, parking lots and gas stations pushed out most other land uses. Further open spaces, parks and farm-land had to make way to new infrastructure and the emerging suburbs surrounding the cities (Register, 2010). Thus todays metropolitan areas don't resemble the traditional towns any-more but rather include hundreds of places what used to be distinct and separate communities in former times (Calthorpe/Fulton, 2001). The author Robert D. Kaplan describes those places as sprawling post-urban pods being bigger than the biggest city but smaller than the smallest county (Kaplan, 1999). With this future vision Kaplan points out one of the major problems urban areas, all around the globe, are experiencing. Sprawl has the habit to fractures local communities, empty the cities and take up the agrarian and natural areas around the traditional towns. With it problems such as crime, failing transportation systems, lack of open space, pollution, and decaying neighborhoods occur. According to Calthorpe and Fulton, however, especially healthy neighborhoods are essential to the well-being of their residents. While those neighborhoods allow different kinds of people and activities to mix up and interact in close proximity, they simultaneously foster a sense of community between them (Calthorpe/Fulton, 2001). Urbanist Jane Jacobs reinforces this argument by emphasizing that strong neighborhoods can form the basis not only for a healthy social life for individuals or the whole community but also for a successful economy (Jacobs, 1985).

Unfortunately modern planning often fails to create the fundamental elements necessary for a community-supporting neighborhood including walkable streets, human-scaled blocks, and usable public spaces like parks, town squares, main streets, and plazas (Calthorpe/Fulton, 2001). The result is a loss of a sense of place and the loss of community-ties, where neighbors know each other, interact and communicate on a daily basis.

In addition to that, Gorgolewski, Komisar and Nasr stress the point that neighborhoods devolve into food deserts where nutritious and affordable food is difficult to get. Originally food availability was a motive power in the formation of former human settlements and the production was closely tied to it. Nowadays, however, the distance between food sources and its market has increased dramatically due to industrialization, affordable transport and new food preservation technologies. Spaces for food production seem to have disconnected from urban design and urban policy. While agriculture became oriented to global markets, zoning commissions started to change formerly to agriculture designated land into profitable districts for future development. Thus the transformation not only promotes sprawl but also neglects the huge potential of urban spaces in contributing to their residents' food supply and the local food security. Equally worse it contributes to the fact that food is being transported over hundreds of miles while loosing its freshness and nutritional value due to polluting transportation and the processing. Over and above that the alteration of the food supply system is connected to the worlds major problems such as pollution, climate change, global poverty, obesity and energy supply, which makes a locally based food systems inevitable in the future (Gorgolewski/ Komisar/Nasr, 2011).

Looking at the described transformation of society and the worsening of urban conditions, it is not surprising that people have always created their own green spaces in order to enhance food security and make their neighborhoods more vibrant and sustainable.

B. Characteristics of Community Gardens

According to Hou, Johnson and Lawson (2009), "the term "Community Garden" is probably best understood as a defined area of tillable land made available to groups of individuals, households, classes, and others to garden" (p.11). Depending on the social and economical situations, there have been different kinds of garden projects such as the vacant-lot garden, School Gardens, Depression-era relief gardens, War gardens, Victory gardens and current community gardens, which can be seen as neighborhood urban garden, children's garden, horticultural therapy gardens, and entrepreneurial job-training gardens (Lawson, 2005). Some of those projects gain for neighborhood improvement and civic activism, while others put focus on job training and environmental education and still others concentrate on the basic gardening for nutrition. Nevertheless, all share the same purpose by supplying diverse people a shared place to meet up and participate in the care of a garden (Hou/Johnson/Lawson 2009).

In the following the historical background of the divers Urban Gardens will be illuminated. Further the specific reasons for their appearance will be explained whereas a particular attention is drawn to the benefits it brings to the individuals and the community of each movement.

I. The History of Community Gardens

The first modern Community Garden (CG) was built in England in the eighteenth century. England at that time was the world's first modern nation with a great transformation of its rural and urban environment. Whereas fenced commercial farms dominated the countryside, the urban picture consisted of narrow workers' houses with no open spaces or gardens surrounding them. Before that time city dwellers cultivated the open spaces within the towns as gardens. Those open spaces not only helped to provide nutrition, but also helped to balance the bad drainage and coal-fouled air that was common and problematic at that time. Unfortunately, Warner emphasizes, those benefits of the gardens were not enough appreciated, which is why almost all remaining open spaces were sold to developers and replaced by buildings with no gardens. This transformation refused many people access to the land they depended on for nutrition and pleasure. This refusal than let to a new social movement: the "self-help" movement, where residents joint to rent parcels at the edge of the city to grow food there. When those got sold for railways, roads and housing developments too, dwellers had to find land further outside the city, which made the venture often impossible (Warner, 1987). Only after a century this movement led into a national policy for the municipal provision of land for the today known: Community Garden (Warner, 1987).

In America the food shortage in World War I led to the so-called "War Gardens" (Lawson, 2005). In order to cope with rationing and high prices, city dwellers supported the domestic food supply in the way that they maintained CG in the vacant spaces of their hometowns. In that sense more food could be send overseas. Later the national organization "National War Garden Committee" took over the task to embolden residents to grow plants. Patriotic slogans such as "Sow the Seeds of Victory" encouraged the planting of many urban gardens between 1917-1919 (Warner, 1987). During the Great Depressing in the 1930s, almost 23 million households took part at several garden programs and achieved to grow produces that were valued at $36 million.

Later on during World War II many brought back the War Gardens of World War I. Back than people started participating in the Victory Garden Campaign to cultivate own food for consumption, recreation, and morale (Lawson, 2005). At that time 44 percent of the vegetables produced in the United States (US) came from the 20 million victory gardens across the country (Warner, 1987). In the 1970 some schools and community garden projects remained, inter alia to connect residents of an area in times of social unrest (Lawson, 2005).

Examples of CG such as Victory Gardens or current models like Havana, where residents grow huge amounts of own food, show that local food production can replace large parts of importing produces form distance places. Nevertheless agribusiness and commercial food systems led to the evaporation of CG after World War II. This discouraged planners to include spaces for CG in their master plans by making it a valid use of land.

In current times, however, urban gardening and urban agriculture slowly gain in importance due to its many benefits to individuals and communities (Gorgolewski/ Komisar/Nasr, 2011). Hou, Johnson and Lawson also made the observation that CG have returned to urban landscapes in current times, since their value to the ecologies of cities are more acknowledged and understood. They even go so far as to argue that CG shifted from being temporary spaces to long-lasting places that include community engagement and even professional design (Hou/Johnson/Lawson 2009).

II. Reasons for the Establishment of Community Gardens

As outlined above different garden projects emerged in the past decades, depending on the social and economical situations at the given time. For each period of time different educational, social or economic reasons played a role why CG had been established (Lawson, 2005).

Lawson subdivides the reasons why CG emerged in the first place and maintained over time, into three main themes. First she emphasizes the importance of CG as a green physical and emotional oasis in a concrete-dominated urban environment to not only the operators but also to other city dwellers who enjoy viewing them. In this sense "nature-in-the-city" is a counterbalance to urban stresses and represent a bridge to the countryside and agrarian lifestyle. As a second theme Lowson highlights the educational role of CG that appeared with the introduction of school gardens in the 1890s to 1920s. It is not only about how to maintain an urban garden, but also to learn about the benefits of teamwork and the importance of patience and commitment. Both, learning how to work and see the rewards of one's labor serves as the basic for future employments. The third theme Lawson talks about is "self-help", where CG are low-cost and direct means to receive food. In this sense, CG turn into social open spaces and bring diverse people, from varied ethic backgrounds, together to work for a common interest. They take matters in their own hands in order to improve their situations. For example in times of war or depression, urban gardens offered neighbors the opportunity to work together and to help themselves by obtaining nutrition. Thus, the community helped to solve a national problem by working, instead of waiting for charity (Lawson, 2005). This theme matches with the explanations of Warner, described above.

III. Benefits from Community Gardens

While there was mostly a dominant objective behind each program (e.g. producing food in times of food shortages), there have always been additional benefits that have survived over time (Lawson, 2005). Due to those multiple personal, social, and environmental benefits, it is not surprising that researchers from different professions currently show increased interest in CG and their impacts. According to Hou, Johnson and Lawson (2009), however, "there is no clearinghouse or easily referenced source that bridges the literature or provides information on recent publications about research related to community gardens" (p.21). Nevertheless, even though CG plays a fairly small role in planning and design literature yet, the benefits they have on the health and sustainability of communities, cities and its residents were analyzed within many studies (Hou/Johnson/Lawson 2009).

As described before many CG are established due to the reason of growing food for consumption or flowers to simply enjoy "urban nature" within the city. A resulting benefit is that people could grow healthy food that may be difficult to obtain otherwise or food that is important to the residents because of their ethnic or cultural heritage. Due to their engagement in urban ecology and conservation (e.g. organic farming, composting, reuse of materials etc.) some CG serve to educate and inspire visitors. For others being separated from daily urban stressors such as traffic and noise, the process of gardening contributes to their well-being and even has healing effects. Very important elements for the physical recreation and emotional restorations of individuals tend to be the proximity to nature, the enjoyment of leisure and the interaction with other people. The health benefits of CG has e.g. been documented in Toronto within a study called "*Growing Urban Health: Community Gardening in South-East Toronto*". The results show that participants experienced better nutrition due to the fresh and healthy vegetables, improved mental health, a sense of community cohesion, and increased physical activity (Hou/Johnson/Lawson 2009).

Further it supports the Community aspect, where different participants take responsibility to maintain a garden together. In addition in holds the benefit of joining diverse groups of people that would not have interacted otherwise and helps them to build up social networks and organize themselves. Spill over effects of the community aspect have led to other local activities such as efforts of neighborhood revitalization or as simple as street cleanups in neighborhoods (Hou/Johnson/Lawson 2009). At the University of California's Cooperative Extension researchers found out, that CG not only beautify urban areas and provide a space where people can gather, but also play an important role in reducing neighborhood crime (Gorgolewski/ Komisar/Nasr, 2011). Jane Jacobs calls the crime-prevention affect that residents have when they watch over space: „eyes on the street". Activating underutilized urban space and promoting community engagement can keep crime away from neighborhoods (Jacobs, 1961).

The educational and informal purpose of many gardens has managed to draw attention to subjects such as gardening, recycling, composting and nutrition. Further the idea of intergenerational activities and neighborhoods outreach arose from those educational programs. Particularly ex-convicts, at-risk teenagers and homeless could benefit from the initiative by learning practical and social skills such as teamwork and responsibility that may be transferable into employment.

Since the 1980s sustainability became a major subject in the field of urban planning. The topic addresses a combination of social, environmental and economical issues (3 E's) in order to enhance conditions for future generations (Wheeler/Beatley, 2008). Since CG deal with all three topics, they play an important role in urban sustainability. On the one hand the nature-in-the-city aspect allows increasing biodiversity in urban environments. On the other hand CG provide green spaces for people to socialize, work together, learn new skills, interact and meet everyday needs. Beyond the neighborhood participation encourage cultural expression and kick off efforts in further community development (Hou/Johnson/Lawson, 2009).

Beyond Gorgolewski, Komisar and Nasr stresses its todays benefits of enhancing food security, reducing the ecological footprint by greening cities and thus improving water management, air quality and encouraging civic activism (Gorgolewski/ Komisar/Nasr, 2011). Lawson on the other hand talks about benefits such as building moral, fighting rising prices and improving nutrition in times of depression. She also points out the utility of bringing people, in particular children, closer to "nature" by reintroducing agrarian values to the urban environment, which has an positive affect on the social and psychological health of urban dwellers and communities (Lawson, 2005).

This rough overview illustrates how many beneficial outcomes CG have on several urban issues and how they serve concerns such as food supply, social justice and education. Nevertheless, urban gardens may also have unwanted effects like noise, smell, and vehicle traffic. In particular in winter time urban gardens can look unpleasant. Thus, careful planning and zoning is necessary to avoid trouble with other parties and to fruitfully integrate CG into urban environments.

2. Possible Urban Spaces to integrate Community Gardens

If there is no coordinated program to implement CG in the urban environment, small programs within different zones might be implemented piecemeal.

Places like city centers often contain a high density where buildings itself might serve as a production area. Often components for food production are planned with the development of new buildings. Former industrial areas offer enough space for urban gardening. In particular greenhouses, raised-bed gardening or rooftop food production works well on space of prior soil contamination and sometimes in urban residential areas (Gorgolewski/Komisar/Nasr, 2011). In particular housetop gardening is one of the oldest forms of gardening. A famous example of ancient rooftop gardens is the hanging gardens of Nebuchadnezzar's. The Greeks also showed early interest in gardens on the their housetops. Taking the ancients as prime example today one can find gardens on apartment buildings, on the roof of garages as well as on sun decks or terraces. Often today's architects even design buildings only with rooftop gardens so that the growing population has green spaces for leisure or for growing trees, vegetables, fruits and flowers in the middle of their neighborhoods. E.g. some rooftop gardens in New York and London contain full-grown trees, waterfalls, fountains, statuary and other pleasant features. The Leadership in Energy and Environmental Design (LEED) certified building of the California Academy of Science in San Francisco as another example implemented a green rooftop that "keeps the building's interior an average of 10 degrees cooler than a standard roof would. The plants also transform carbon dioxide into oxygen, capture rainwater, and reduce energy needs for heating and cooling" (California Academy of Science, 2013). Despite the many innovative ideas, there are still many bare roof spaces in congested areas. However, the more typical black rooftops there are, the more heat is trapped within the area, which leads to "urban heat islands". Instead, those drab areas could be ideal to become a green space to beautify the area and improve water management and air quality as well as the well-being of participants by encouraging civic activism in rooftop gardening (Brett/Grant, 1967).

Commercial areas with great parking lots and lawns as well as highways, railways or power line corridors that have enough adjacent wasteland can be used for CG. Older suburbs with larger and sometimes vacant lots on the one hand provide land for growing food and flowers. Lower-density suburbs that are newly designed, on the other hand, can use public land to integrate CG. Front and back yards play a big role in both cases (Gorgolewski/Komisar/Nasr, 2011). Especially interesting land for growing food are the brownfields within a city such as former military, infrastructure or traffic areas. Often those sites are not developed yet and still wait for a long-term development. As shown in Figure 2 they provide good short-term possibilities for CG and many benefits like offering a low-cost food source, fostering community relationships, and cleaning up vacant lots.

[See Appendix Figure 2]

CG Communities never have a lot of funds that's why cost-intensive restructuring of the areas aren't possible. Therefore CG have to adapt to the conditions given at every site. Since they CG are very flexible and creative they can even survive on contaminated sites by means of portable garden boxes and containers. To show how and where urban gardening nowadays take place, two examples will be provided in the next section

A. International Case Study: "Prinzessinengärten" in Berlin, Germany

Berlin is a prime example of urban gardening movement in Germany. One of the nation-wide known CG is the "Prinzessinengärten". In 2009 "Nomadisch Grün", a non-profit organization, rented a 6.000m^2 area from the city. The land they created the urban gardens on had been wasteland for about 60 years without any concrete development intentions for the future. The idea to implement a CG close to the Moritzplatz in the middle of Berlin came due to one of the founders' travels to Cuba, where he experienced the benefits of the well-known urban gardens of Havanna. Along with other people of the neighborhood and their friends, they cleaned up the lot and built transportable organic vegetable plots and opened a garden café. As shown in Figure 3 the former vacant and unappealing lot turned into a green oasis in the middle of the urban environment of Kreuzberg, Berlin.

[See Appendix Figure 3]

The general idea of the founders was to implement mobile gardening and transform unused spaces like car parks, building sites and rooftops temporarily in CG. Since the available lots are usually either contaminated or paved and might be developed with time, the gardens are mobile. This method makes gardening independent from the ground below and allows organic farming (Clausen/Müller-Frank, 2012). The specific aim of Prinzessinengärten is to produce fresh and healthy food for the neighborhood, increase biodiversity, decrease CO^2 and improve the microclimate. Beyond, the participants hope to exchange knowledge and competencies, make residents aware of a sustainable lifestyle due to education and above all foster a sense of community within different groups of people. According to Nomadisch Grün (2014): "In an unobtrusive and pragmatic way, such gardens raise the question of how we want to live in our cities in the future" (Prinzessinengärten)." While Prinzessinengärten got its inspiration from Cuba and North America, the CG today serves as a role model for all kinds of garden projects within Germany.

B. Local Case Study: Smart Farm San Diego

Many CG have emerged in San Diego City over time. The San Diego Community Garden Network (SDCGN) is one of the approaches to creating more green spaces in the city. Looking at their registered CG within this network (see Figure 4), it is clear that the demand for urban grading is present and growing. Their mission is to: "help create, support and grow community gardens that enrich their neighborhoods by enhancing food security, promoting a sustainable environment and fostering community based educational opportunities and community building" (SDCGN, 2011). One of the community gardens listed at SDCGN is "Smart Farm", located at the corner of F Street and 15th in the "Makers Quarter".

[See Appendix Figure 4]

Like Prinzessinengärten in Berlin, the Smart Farm has transformed a vacant lot of San Diego's East Village into a green Community Garden and a center for learning about nature. Figure 5 shows the area before and after the introduction of CG at the lot.

[See Appendix Figure 5]

Susan Madden Lankford, a San Diego-area photographer, author, and activist founded the Human Smarts Non-profit organization in 2012. With her project she wants to provide opportunities for creativity and self-sufficiency by offering programs in photography and other arts, gardening, and academic enrichment. The main goals of the Smart Farm is to help underserved young people (age 5 to 15 years) to develop respect for nature and the environment, provide opportunities for inquiry, observation, and experimentation, teach budding gardeners to nurture and care for other living things, motivate picky eaters to appreciate fruits and vegetables, and promote physical activity and positive outdoor experiences. With its "Gardening and Arts program", the project stimulates children to connect with nature and develop creativity. With the day-program "Craft Day at the Farm" for example, children learn how to paint rocks and faces, and get animated to put seeds and plants in the soil of the Smarts Farm garden plots (e.g. 10th April 2014, see Figure 1 below). Further there is the opportunity to lease easy-to-garden planter boxes for residents of the community. Special programs such as "Bugs, Soils and Seeds" and "Garden Exploration Day" teaches new gardeners of the community the importance of growing food locally while master gardeners share their expertise on how to do so (Humane Smarts, 2013). Beyond, the so-called "Museum of Outdoor Photographic Art" (MOOPA) is a project in collaboration with Smart Farm that has enhanced the quality of the streetscape design. By transforming an empty warehouse at F Street, between 14th and 15th Street, into an impressive 386 feet long ArtWall, the surrounding neighborhood has gained in creativity and beauty. The installation shows work of three noted photographers and has become a point of interest to many people (Humane Smarts, 2013).

In that way the Smart Farm doesn't only beautify the area by greening the vacant lot but also offers a place for diverse community members to interact and enhance their sustainable living patterns. Beyond it aims to teach the youth but also the adult Smart Farm Community in matters of sustainability, food production and respect for nature and living organisms. With their diverse range of events and activities they enliven the area and foster community interaction.

Figure 1: Children's Program at Smart Farm San Diego
Source: Pictures by Theresa Hayessen, San Diego April 10th, 2014

3. Role of Urban Gardens in the Field of City Planning

Looking at the challenges of todays' twenty-first century cities, it becomes clear that creating sustainable human communities will be necessary. Urban environments have to become a climate-friendly, pleasant place to live, where every care is taken to conserve the natural resources. The broad field of "City Planning" can play an important role in meeting these objectives. To understand how CG can serve as a tool for city planning, it is important to have a look at former and current planning visions of gardening within the urban context.

A. Roots of Urban Planning

In the late nineteenth century the Industrial Revolution let to an extremely rapid growth of cities. Many urban problems such as inadequate sanitation, water supply, lack of transportation and housing with adequate light, water, and sewer facilities occurred within the cities. In that time the organized profession of urban planning came into action to overcome those problems. Architects e.g. proposed housing regulations to ensure access to light and air and landscape architects started the Park Movement to beautify the city. Visionaries such as Ebenezer Howard, Patrick Gedds, and Ildefons Cerda came up with holistic ideas to overcome those urban problems (Wheeler, 2013).

In Germany the planning tool of zoning emerged in cities such as Frankfurt am Main in 1890s. In Britain and North America city planning as an organized profession become established in the 1910s and 1920s. Back than municipalities first created planning commissions, planning staffs, and zoning regulations. In 1909 the University of Harvard established the first US academic city-planning course. Many other universities followed their example in the following years and let to the favor of more pragmatic approaches than those of Howard and Geddes. However, many problems of today's urban problems were caused through the abundance of holistic ideas. Modernist planning e.g. bulldozed neighborhoods of older housing in favor of sterile new apartment blocks that failed the pressing needs for equity and improvement in social conditions (Wheeler, 2013).

In the second half of the twentieth century every city in the US had created a planning department composed of the "current planning division" that is responsible for development permitting and the zoning code, and an "advanced planning division" that oversees the long-range development. " According to Wheeler (2013), "planners also came to work within many other types of government agencies – regional agencies, state and national governments, international development agencies, transportation authorities, park districts, and public utilities among them – as well as for private corporations, development companies, consulting firms, community development corporation, and various types of nonprofit nongovernmental organizations (NGOs)" (p.15). The interaction with many different fields shows the network and crucial control planners have on the urban development.

B. Planning in Relation to the Idea of Urban Gardening

As mentioned earlier the garden city model of Howard (1898) made an early and important contribution to city planning in response to the poor housing and living conditions as well as the horrendous rising land prices in the fast-growing cities. Howard planned Garden Cities as economically independent neighborhoods with greenbelts surrounding them. The strict separation of town and country should be repealed in order to avoid the disadvantages of the city and maintain its advantages. Beyond, areas of residences, industry and agriculture within the garden city were separately arranged, which made reliable transportation essential. The idea of garden cities also influenced U.S. city planning. Examples are Queens in New York or Woodbourne in Boston. In most cases, however, Howards' idea was only implemented by means of garden suburbs instead of complete garden cities (Meacham, 1999).

In the late 1920s Frank Lloyed Wright, a U.S. architect, developed the "Broadacre City". This urban concept intended residents to live in a suburban, decentralized landscape.

Each family would be given one acre of land in order to farm their own land and thus live independently. He assumed that every modern man had the right to own a car and use it as the most important means of transportation. Looking at today's common "Edge Cities" in the U.S. they act as an incomplete version of the Broadacre City (Fishman, 1982). The concept of including agriculture and green spaces in urban landscapes as an alternative to parks and plazas emerged in 1980s in the U.S.

Regarding today's urban issues it is no wonder that developing sustainable urban settlements, like urban gardens, gain increased interest again. Especially the fact that food-related initiatives can fight some current urban problems by reducing the heat-island effect in sealed urban areas, lower water treatment costs and cut health costs due to better nutrition of participants, political attention has grown. Driven by this increasing interest and the need for sustainable cities, urban designers and planners, as well as architects are adding an urban food production focus onto their work. According to Gorgolewski, Komisar and Nasr (2011), "a truly sustainable urban community can only be created, however, when the benefits and implication of urban food production, including appropriate building design, urban spaces, and components that can contribute to the local food supply are addressed in a pragmatic way (p.14)." To do so, it does not necessarily need professional planning. Nevertheless, they can be of great help to find or make spaces available within the city where CG can be realized. Especially the new generation of students, that grows up with the awareness about sustainability that their parents' generations lacked, should see urban gardening as a point of departure. Integrating urban food production as an integral part of cities overarching plan simultaneously enhance sustainability by greening and beautifying the urban environment, increase the access to fresh produces, providing a sense community, and designing buildings and landscapes in an sustainable way. Beyond, programs such as LEED reflect the increased emphasis on sustainable planning and design in todays planning affords. Due to those approaches topics like green roofs or living walls have gained attention and could provide more space for community gardening programs (Gorgolewski/Komisar/Nasr, 2011).

Over time planners have been successful to meet some urgent goals in urban development, but it hasn't been enough in other areas. Beyond today they are facing new problems such as climate change, excessive reliance on motor vehicles, resource depletion and inequality, that makes sustainable visions in cities even more indispensable. With the aim of food and energy supply, beautifying the urban landscapes, and bringing the sense of community back to neighborhoods CG addresses many current urban issues. This emphasizes the fact that CG, next to other ideas, has become an essential urban planning and design tool in order to mitigate and reduce some of the major problems todays' cities are facing (Register, 2006).

C. How to implement Urban Gardens in the Field of City Planning

The history of planning, including the different approaches, has shown that the longing for nature in the city has existed for a long time. In regard to the listed benefits it is obvious that CG can do much more than just to increase the value of land (e.g. enhancing vacant lots). They have rather medium- and long-range positive structural, economical, and social effects on communities. Thus, on the long run they offer a great potential for a sustainable and healthy city development with livable, equitable and ecological communities.

In the following some methods will be presented on how to implement the tool of CG in the field of city planning in order to draw all potential benefits from them.

I. Creating "Agriculture Incentive Zones" within Cities

The biggest challenge garden projects are facing is to find available space within the urban environment. Especially in cities where land and property values are high, it is difficult to find an appropriate area. This, however, is mostly the case metropolitan areas. Thus, land tenure represents a major obstacle for the financially hardly competitive garden projects. Cities and municipalities can overcome this problem by removing specific plots of land from the free market and dedicate them to special projects such as urban gardening.

Another tool to encourage urban gardening and small-scale farms in urban pockets is to implement "agriculture incentive zones". According to Romney (2013) "the concept for the zones is a hybrid of the Wiliamson Act, which offers tax subsidies to owners of rural land maintained for agricultural purposes, and the Mills Act, under which cities may enter into contracts with private owners who receive subsidies in exchange for restoring and preserving historic buildings." California cities and counties now have this novel tool at their disposal and can apply it on voluntary basis. The Bill No. 551 (approved September 28, 2013) is a market incentive for property owners who want to wait to develop their land. According to the California Legislative Information (2013): "This bill would enact the Urban Agriculture Incentive Zones Act and would authorize, under specified conditions and until January 1, 2019, a city, county, or city and county and a landowner to enter into a contract to enforceable restrict the use of vacant, unimproved, or otherwise blighted lands for small-scale production of agricultural crops and animal husbandry." This law brings advantages equally to the landowner in matters of tax benefits, to the urban gardener searching for a plot to work on and the community since vacant land is being used in a productive way.

II. Vacant lot Asset Mapping

Even in cities where property values are not as high and more land is available in terms of brownfields and vacant lots, it is not easy for urban garden projects to find obtainable plots. That is because garden communities often lack information and contacts. Municipalities could support them bridging this problem by delivering all necessary data on vacant lots. Interesting for that task could be the idea of the Brownfield/Vacant lot Map, which was created by Urban Studies and Planning Students from the University of California San Diego and State University San Diego in corporation with some community partners. According to the Superfund

Research Center (2012): "The objective is to set up a community-based system to document the type and distribution of vacant properties in San Diego while identifying and cataloging vacant lots that are good sites for community gardens."

[See Appendix Figure 6]

With this type of mapping the cities could give all necessary information to all sorts of groups searching for vacant land and draw benefits from it themselves (e.g. beautifying unappealing areas like Smart Farm did).

III. Vacant Lot Cadaster

Another possibility would be the preparation and publication of a brownfield or vacant lot cadaster that lists all possible locations for urban gardens. One way would be that municipalities and cities coordinate the contact between property owners and demanders and provide information only on demand. Another way could be to coordinate the offers and the demand by means of a website in the style of a classic property page (e.g. www.immobilienscout24.de). The owners, including cities and municipalities, can offer their properties along with all information online. The Urban Garden projects have the opportunity to contact the landowner via the webpage in order to obtain more information (Tobisch, 2013).

IV. Integrate Urban Gardening as an Official Instrument of City Planning

Urban gardening becomes more and more important in many cities and the idea has spread to many different countries. However, urban garden projects often have to fight for plots to serve as CG areas. E.g. in the 1990s in New York, when Rudolph Guiliani was Mayor, urban gardeners had to struggle hard to keep their land. He wanted to sell off vacant lots that were being used as CG for more profitable development. If it wasn't for public campaigns these areas could not have been saved. According to Wheeler (2013), "Rather than fitting community garden in haphazardly on leftover scraps of land, municipalities should look on their gardens as an essential strategy for making dense neighborhoods livable (p. 317)." Professionals involved in city and regional planning can play a crucial role in the process of sustainable planning of urban environments. They might be helpful in identifying and acquire surplus city sites for urban garden projects or require developers to include space for gardening within large building projects.

Generally creating many different opportunities for open spaces needs to become a much more integral part of planning. Communities can promote urban gardening through various land use laws while land use policies can secure access to and ensure the preservation of plots for gardening uses. Municipalities should strengthening open space requirements within zoning codes and subdivision regulations. The regulations can ensure that urban gardening occurs in appropriate locations and under proper conditions. Since every neighborhood varies in its characteristics such as population density, resident interest, soil suitability and availability of plots, however, there is no general agricultural land use policy. Communities have to fit the regulations to their special needs and context (nplan, 2012). Specific design guidelines for CG or other open spaces and actively acquiring land or easements for those ideas is necessary. State and federal government on one hand can be of great help by assisting them by means of grant funding and modification to subdivision law requiring additional open space. City governments on the other hand can ensure that those open space are used through urban gardens that not only cultivate the land but also schedule and manage events in these areas (like Smart Farm does in San Diego). This ensures that the spaces are extensively used and that they contribute to the diversification of neighborhood activities, to the sense of community, to crime reduction, and to education (Wheeler, 2013).

Beyond, in order to argue in favor of the establishment of CG, more research on the long-term benefits has to be accomplished and good access to relevant data should be secured.

Further education on the advantages of the gardening projects will help to spread the idea and make officials sensitive to the subject.

Regarding the challenges todays' urban environments are facing it is obvious that more sustainable approaches are needed. The method of urban gardening has proven to have many beneficial effects on cities and their residents. Even tough CG are only one element among others on how to solve urban problems, it is crucial to be aware of how small steps can lead to larger long-range improvements. And although the profession of planning has a key role in the process of sustainable development, preparing for a better future involves all of us. Prime examples like Smart farm in San Diego and Prinzessinengärten in Berlin have shown how individuals used the tool of urban gardening to set free all kinds of benefits for the environment as well as the human beings that are depended on it. More such initiatives have to take place and urban planning should support it by making community gardening to an official and well-used tool of the process.

Appendix

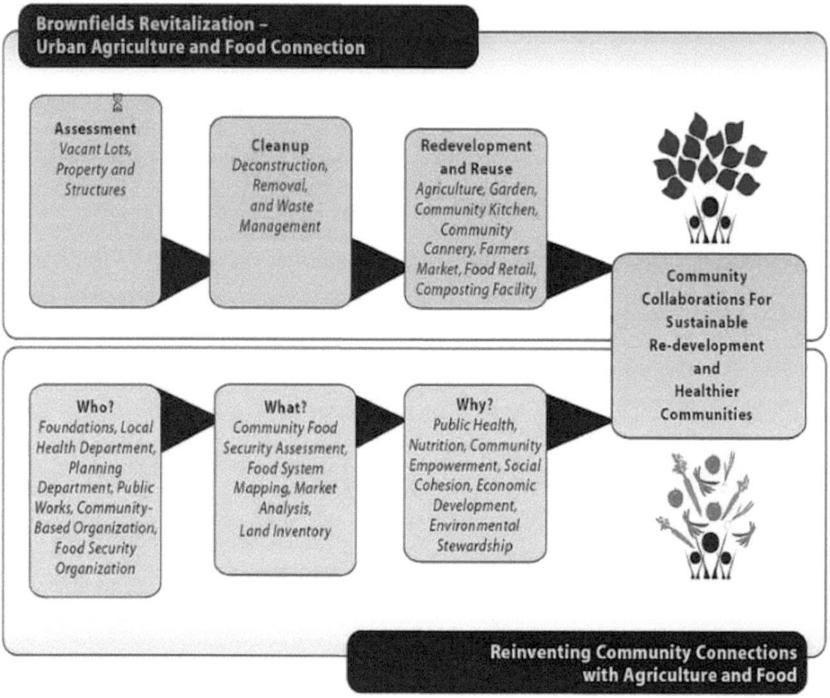

Figure 2: Stimulating productive use of brownfield sites through urban gardens
Source: Superfund Research Center UCSD, 2012

Figure 3: Lot of Prinzessinengärten before and after the implementation of CG
Source: Prinzessinengärten, 2014

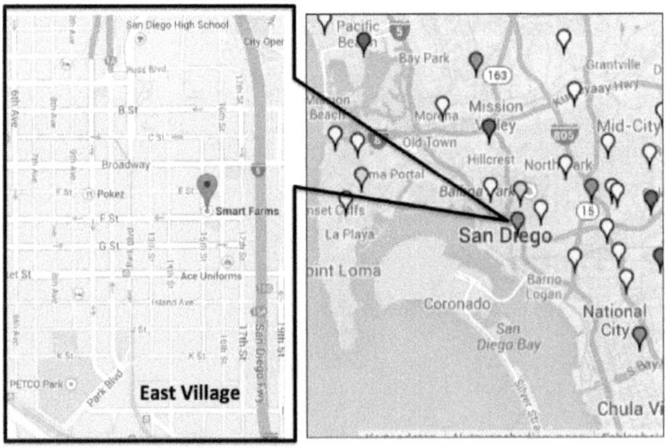

Figure 4: Urban Garden Network and the location of Smart Farm in San Diego Downtown
Source: Own depiction based on http://www.sdcgn.org & Google Maps

Figure 5: Lot of Smart Farm before and after
Source: (L) Humane Smarts, 2013 / (R) Picture by Theresa Hayessen

Figure 6: Brownfield/Vacant lot Asset Mapping
Source: Superfund research Center UCSD, 2012

Bibliography

Books:

Brett, W & Grant, K. *Small City Gardens*: New York, Abelard-Schuman, 1967

Calthorpe, P. & Fulton, W. *The Regional City*: Washington DC, Island Press, 2001

Clausen, M. & Müller-Frank, S. *Prinzessinnengärten: Anders gärtnern in der Stadt*: Leipzig, DuMont, 2012

Fishman, R. *Urban Utopias in the Twentieth Century: Ebenezer Howard, Frank Lloyd Wright, Le Corbusier*: Cambridge, MIT Press, 1982

Gorgolewski, M. & Komisar, J. & Nasr, J. *Carrot City: Creating Places For Urban Agriculture*: New York, Random House, 2011

Hou, J. & Johnson, J. & Lawson, L. *Greening Cities Growing Communities: Learning from Seattle's Urban Community Gardens:* Washington, University of Washington Press, 2005

Jacobs, J. *Cities and the Wealth of Nations: Principles of economic life:* New York, Randon House, 1985

Jacobs, J. *The Death and Life of Great American Cities:* New York, Randon House, 19

Kaplan, R. *An Empire Wilderness: Travels Into America's Future:* New York, Random House, 1999

Lawson L. *City bountiful: A century of community gardening in America:* Berkeley, University of California Press, 2005

Meacham, S. *Regaining Paradise: Englishness and the Early Garden City Movement*: Michigan, Yale University Print, 1999

Register, R. *Ecocities: Rebuilding cities in Balance with Nature:* Gabriola Island, New Society Publishers, 2006

Tobisch, C. *Oasen im Beton; Urban Gardening als Instrument zur Attraktivierung und Belebung von Brachflächen,* Diplomarbeit, Fakultät Raumplanung, Technische Universität Dortmund, 2013

Warner, S. *To dwell is to garden: A History of Bostons's Community Gardens*: Boston, Northeastern University Press, 1987

Wheeler, S. *Planning for Sustainability; Creating livable, equitable and ecological communities,* 2. Edition, Abingdon, Routledge, 2013

Wheeler, S. & Beatley, T. (Edt.) *Sustainable Urban Development Reader*: 2. Edition, New York, Routledge, 2008

Internet Site:

California Academy of Science. (2013) *The Living Roof.* Retrieved March 25, 2014, from: https://www.calacademy.org/academy/building/the_living_roof/

California Legislative Information. (2013) *AB-551 Local government: urban agriculture incentive zones.* Retrieved April 10.2014, from: http://leginfo.legislature.ca.gov/faces/billNavClient.xhtml?bill_id=201320140AB551

Humane Smarts. (2013) *Where Heart meets Mind,* Retrieved March 25, 2014, from: http://humanesmarts.org/mission/

National Policy and Analyses Network. (2012) *Seeding the Cits; Land Use Policies to Promote Urban Agriculture,* Retrieved April 10.2014, from: http://changelabsolutions.org/publications/seeding-city

Prinzessinengärten. (2014) *About Princess Gardens.* Retrieved March 25, 2014, from: http://prinzessinnengarten.net/about/

Romney, L. (2013) *New California law aims to cultivate urban agriculture*. Retrieved April 05.2014, from: http://articles.latimes.com/2013/oct/02/local/la-me-urban-agriculture-law-20131003

San Diego Community Garden Network. (2011) *About SDCGN.* Retrieved March 25.2014, from: http://www.sdcgn.org/about-2/

Superfund Research Center UC San Diego. (2012) *Brownfields/Vacant Lot Asset Mapping*. Retrieved March 25, 2014, from: http://superfund.ucsd.edu/index.php/initiatives/detail/brownfields